Copyright © 2021 Spunky Science

All rights reserved. No part of this book may be altered, reproduced, redistributed, or used in any manner other than its original intent without written permission or copyright owner except for the use of quotation in a book review.

Letter Practice

Aa Bb Cc Dd
Ee Ff Gg Hh
Ii Jj Kk Ll
Mm Nn Oo Pp
Qq Rr Ss Tt
Uu Vv Ww Xx
Yy Zz

LAB RULES

Wash your hands

Use goggles

Do not eat or drink

Be responsible

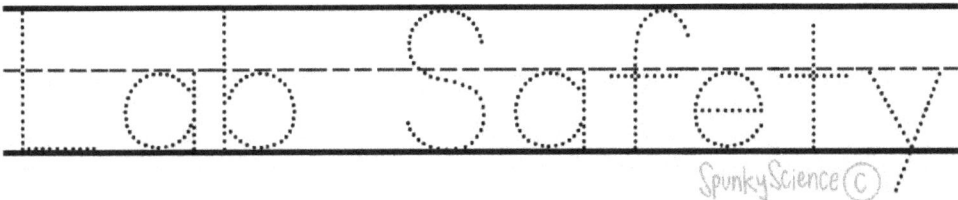
Lab Safety

Spunky Science ©

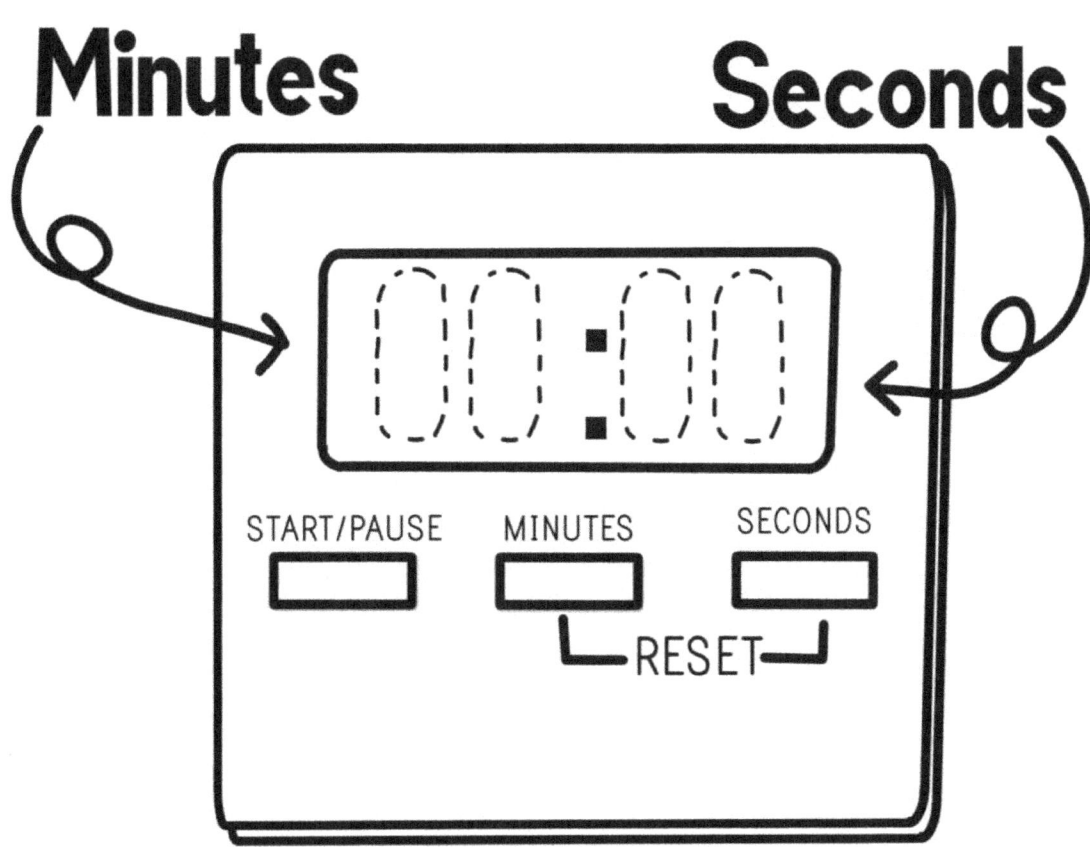

A timer is used to record how long something is happening. **This timer measures time in minutes and seconds and has a pause** button.

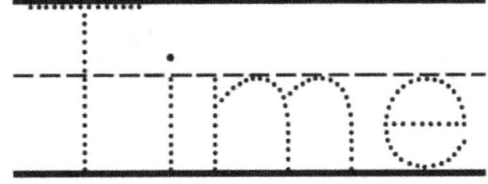

RAIN GAUGE

A rain gauge measures how much rain falls in a specific area.

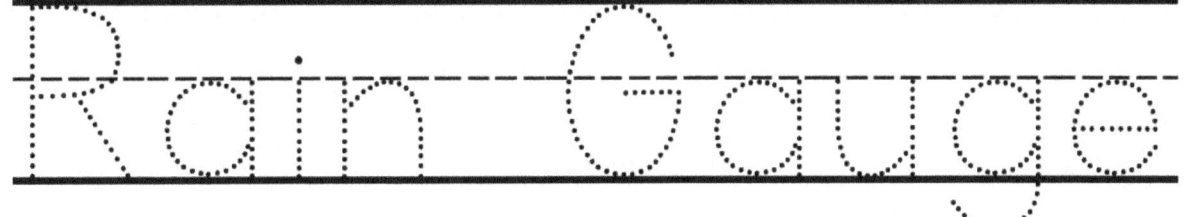

A ruler is a tool that is used to measure the length of an object. Each side measures in a different unit.

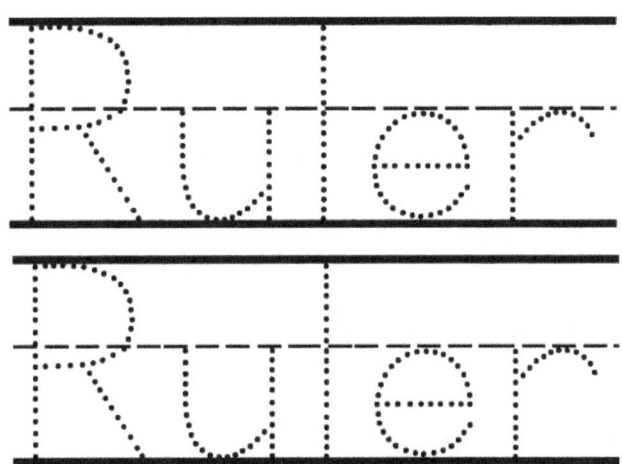

THERMOMETER

A thermometer measures how much heat something has. If it has little heat, the number is lower and if it has a lot of heat energy, then the number is higher.

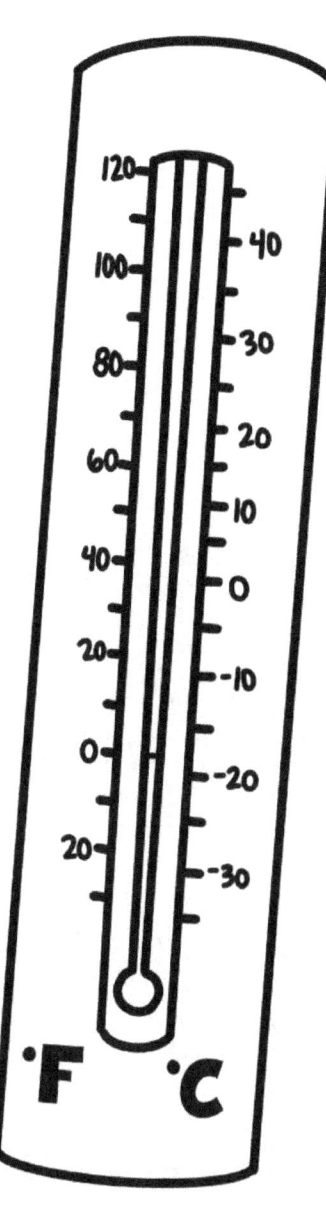

Some tell you the number while others you have to figure out.

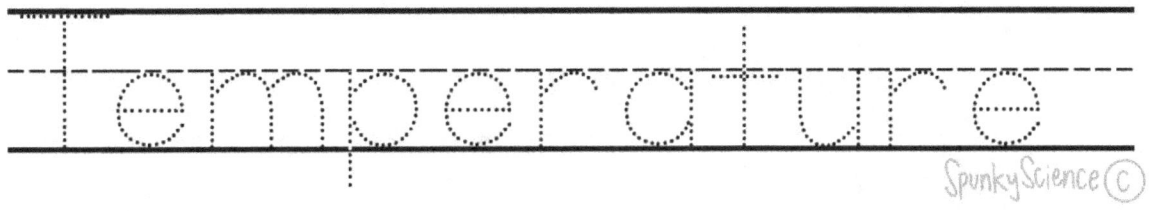

RECYCLING

Recycling is how we take trash and make it into something new. It is good for the environment because it reduces the amount of trash in the landfill.

ENGINEERING DESIGN PROCESS

- Identify the problem
- Brainstorm solutions
- Select a design
- Build a model or prototype
- Test and evaluate
- Optimize the design
- Share the solution

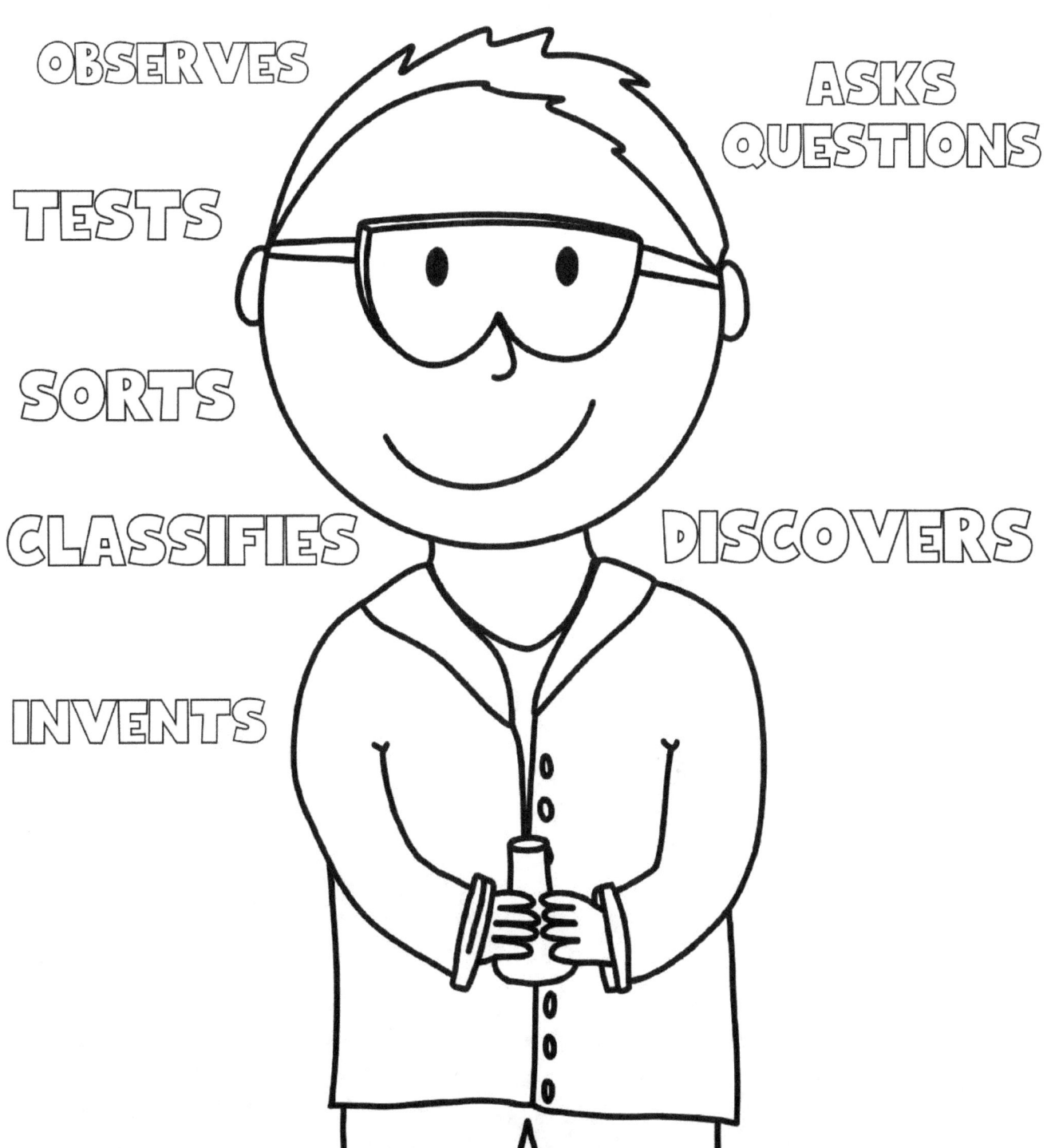

COMPARING IN SCIENCE

Bigger or Smaller

Which piece of toast is larger?

Heavier or Lighter

Which weight is lighter?

Shape

The Pyramids are triangular shaped prisms

Color

What color did the Statue of Liberty change to and what color did it used to be?

Texture

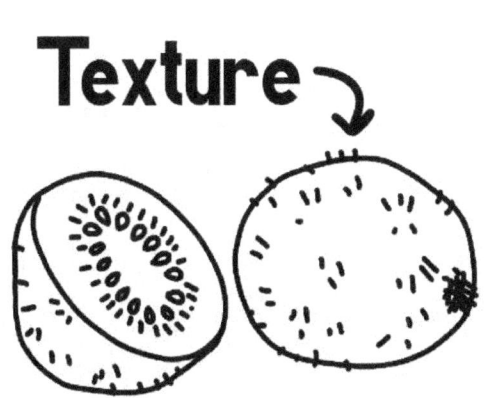

This Kiwi has a fuzzy texture on the outside and a smooth texture on the inside.

PHYSICAL PROPERTIES

Draw objects that you see in the room. You should be able to classify them into the boxes below.

Color
Blue, yellow, red, etc.

Size
Large, medium, small

SHAPE
Round, triangular, square, etc.

TEXTURE
Soft, smooth, rough, hard, bumpy

MATERIALS

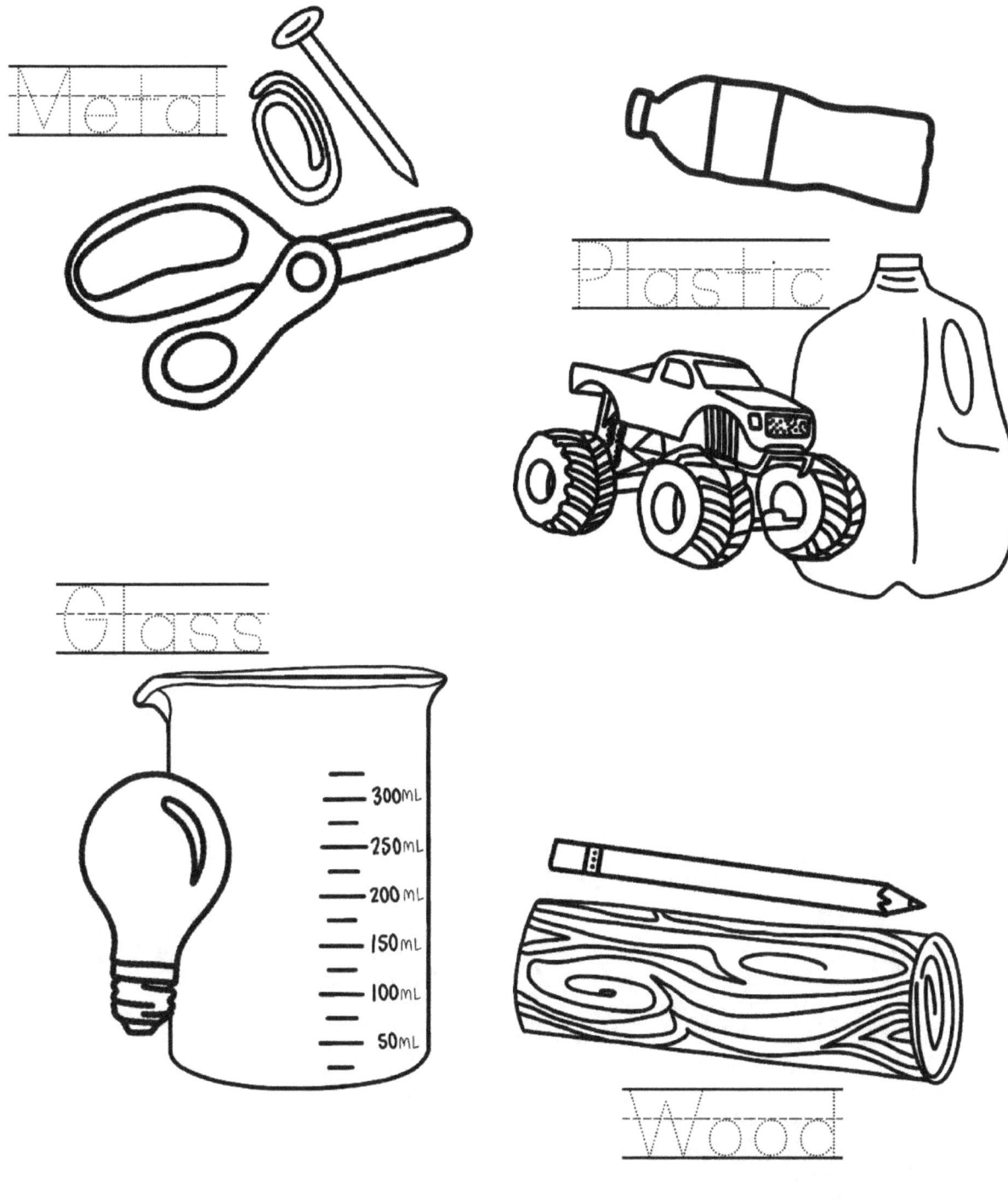

MOVEMENT

Objects can move in all sorts of directions. Draw something that moves in that direction in each box!

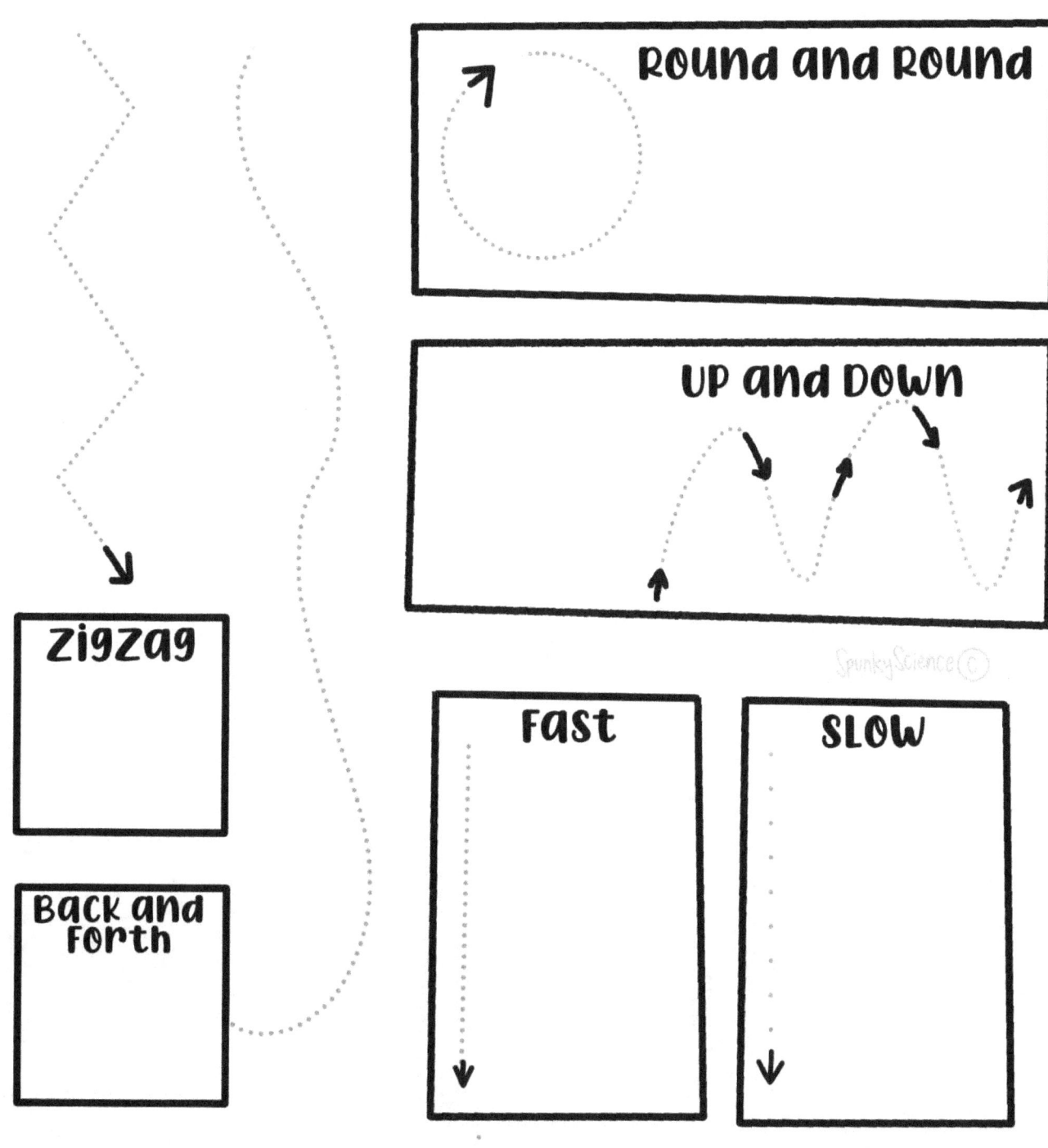

HEATING AND COOLING

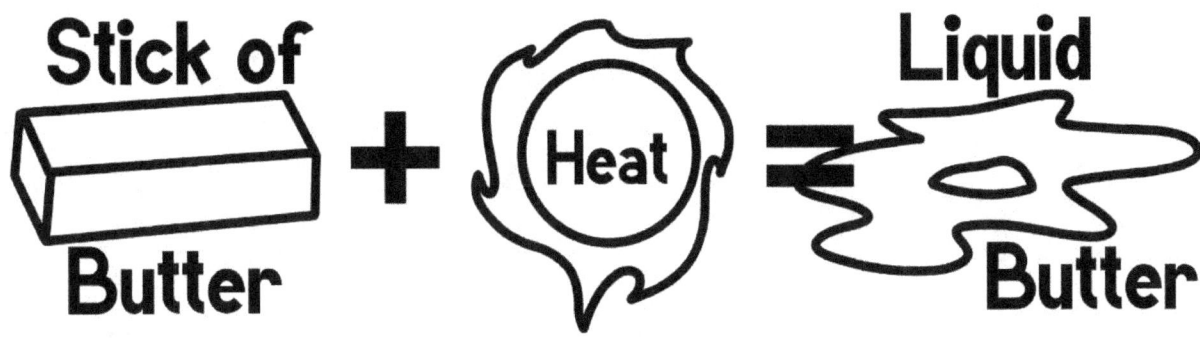

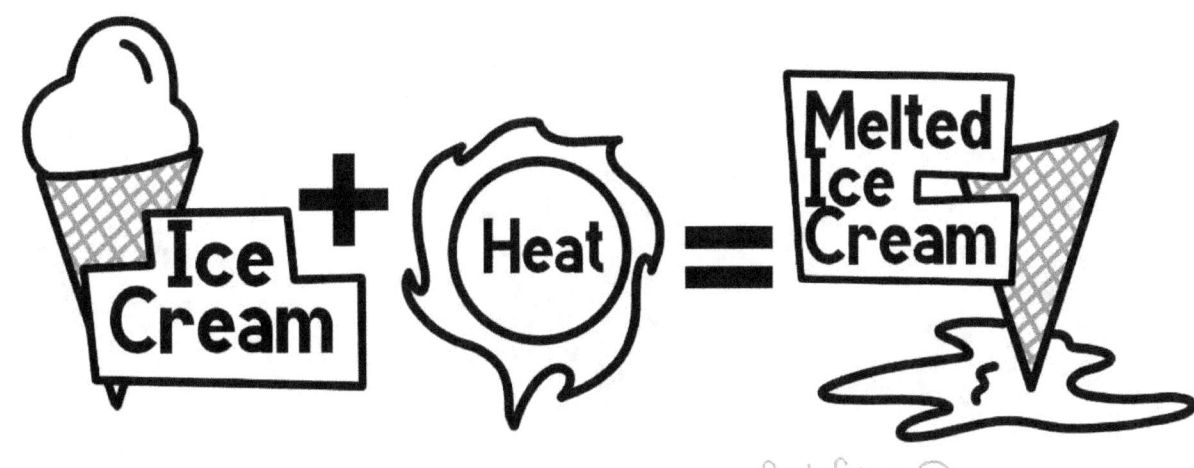

FORCES

Forces are happening all around you! Some of these forces you can see and some you can't see.

Each person pulls in different directions

This person is pushing this cart

Gravity pulls this apple to the ground

All forces are either a push or pull

ENERGY

Light

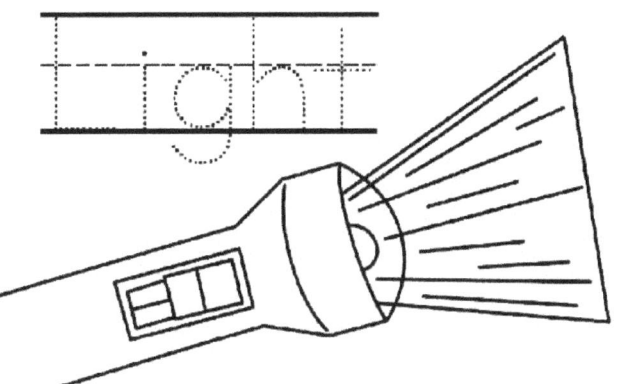

Light energy is the kind of energy that you **can** see.

THERMAL

Thermal energy is the kind of energy that you **can** feel.

SOUND

Sound is caused by vibrations of an object. In a guitar the strings vibrate which causes a sound.

Things that put off their own light

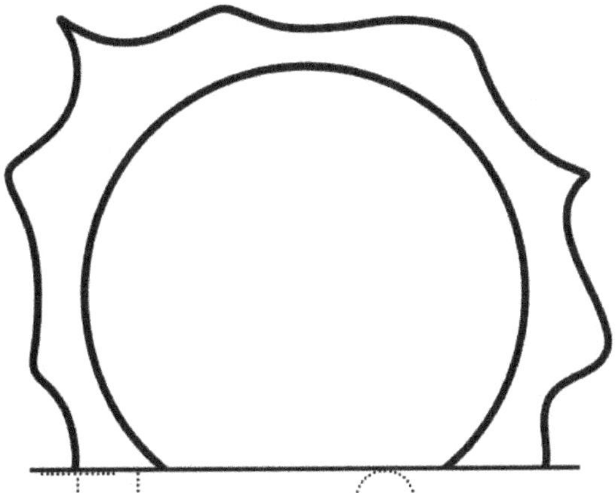

The Sun

Fire

Lightbulb

Little Dipper

Stars

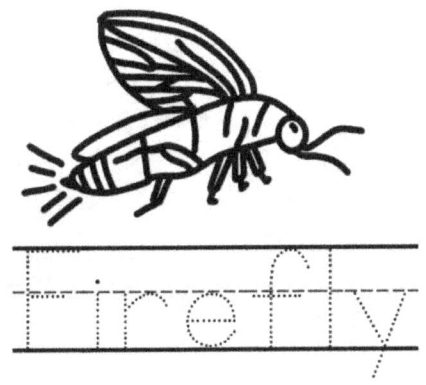
Firefly

This instrument is called a CLARINET.

When the musician blows air into the clarinet, the air causes a piece in the instrument to vibrate. These vibrations produces the clarinets sound.

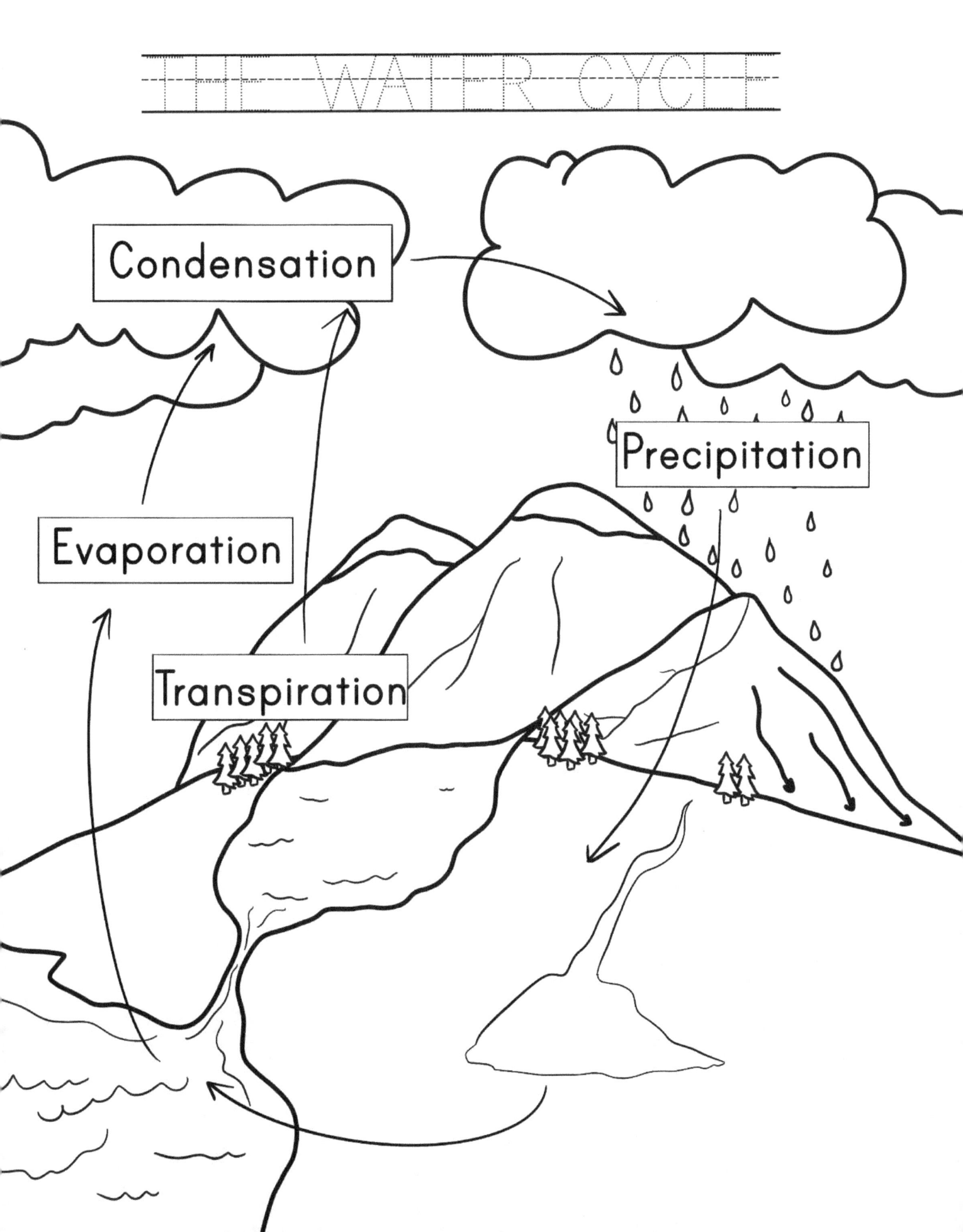

CLASSIFYING SOIL

There are lots of different types of soil. Soils are categories by their size, color, and texture.

TEXTURE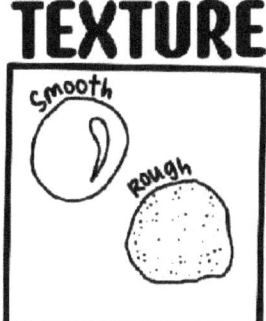

COLOR

SIZE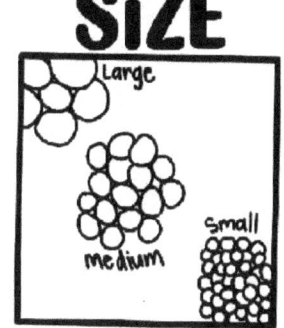

Sand

Draw:

Size of particles:

Texture:

Color:

Loam

Draw:

Size of particles:

Texture:

Color:

Clay

Draw:

Size of particles:

Texture:

Color:

NATURAL WATER SOURCES

A lake is a large body of water with land around it.

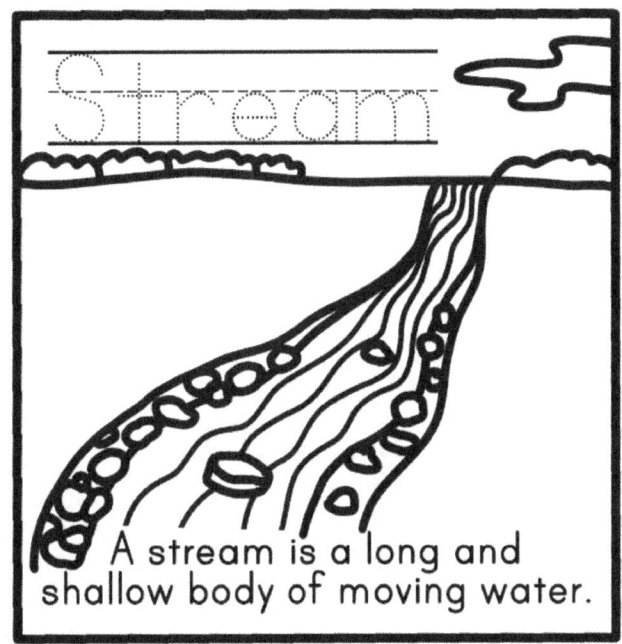

A stream is a long and shallow body of moving water.

An ocean is a very large body of saltwater.

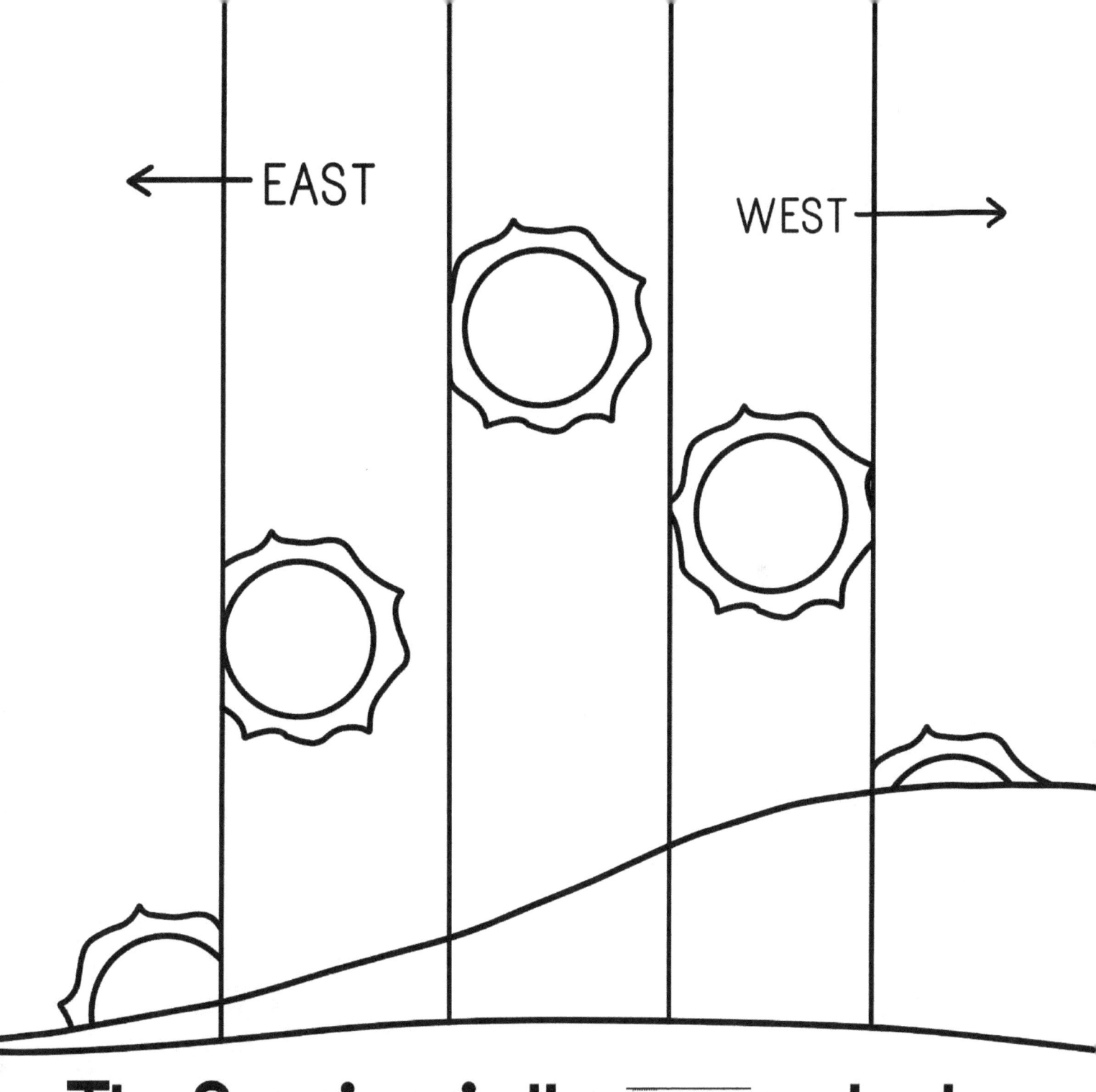

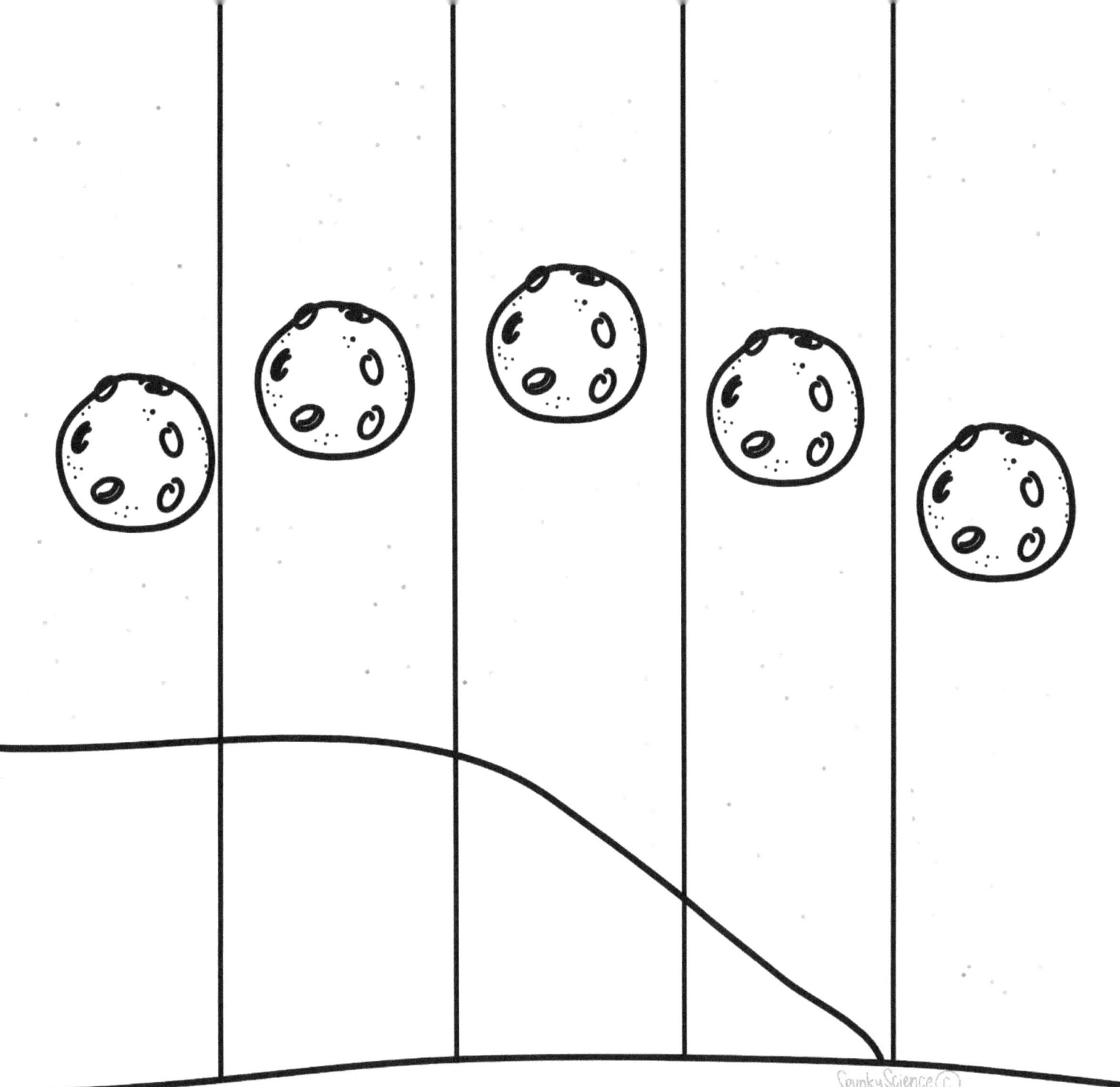

The Moon moves across the sky course of the night as the Earth rotates.

STARS

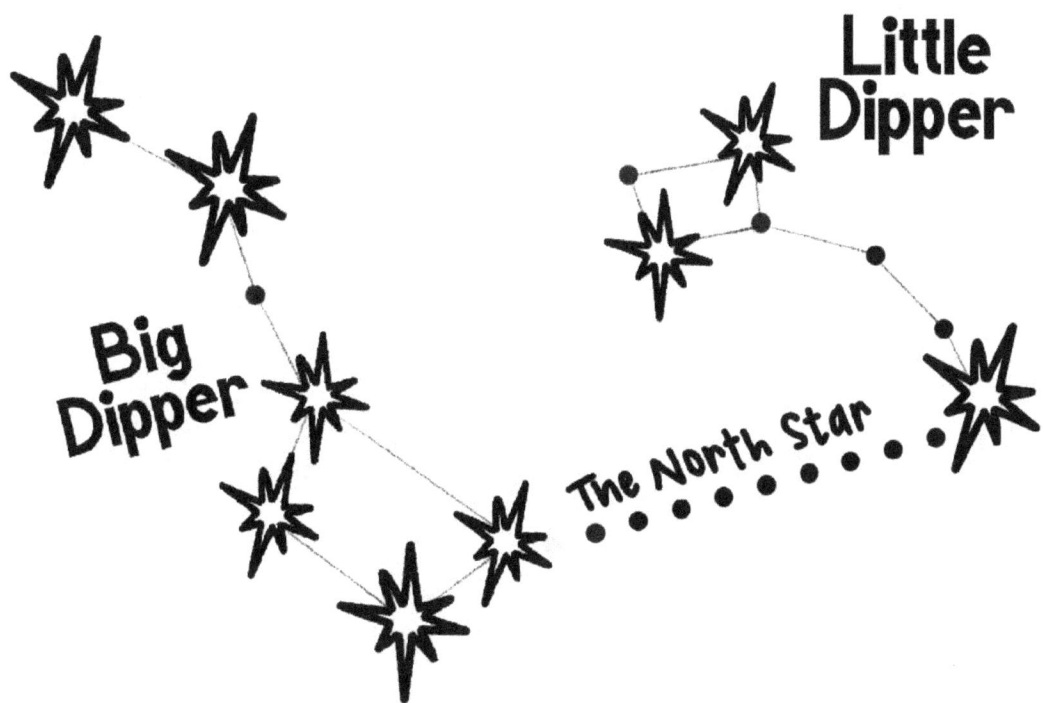

Stars are only visible at night

Let's **figure out the number of** sunny, windy, rainy, **warm days in one** month!

Sunday	Monday	Tuesday	Wednesday	Thursday	Friday	Saturday

Draw these symbols on the days that shows that type of weather.

Sunny — *Cloudy* — *Windy* — *warm*

Let's count! Write the number of days that were sunny, windy, rainy, and warm this month in the spaces below.

____ **Sunny**

____ **Cloudy**

____ **Windy**

____ **Warm**

Spunky Science ©

DAY AND NIGHT

Daytime is when the Sun is shining on your side of the Earth.

Nighttime is when the Sun is on the other side of the Earth, so its light and heat do not reach you.

LIVING THINGS

Living things have needs in order to survive where non living things do not.

NON LIVING THINGS

ECOSYSTEM

A terrarium is a closed environment of living things that has everything they need to survive. This ecosystem depends on the right amount of water, sunlight, nutrients, and soil.

Living things need both living and non-living things to survive

FOOD CHAINS

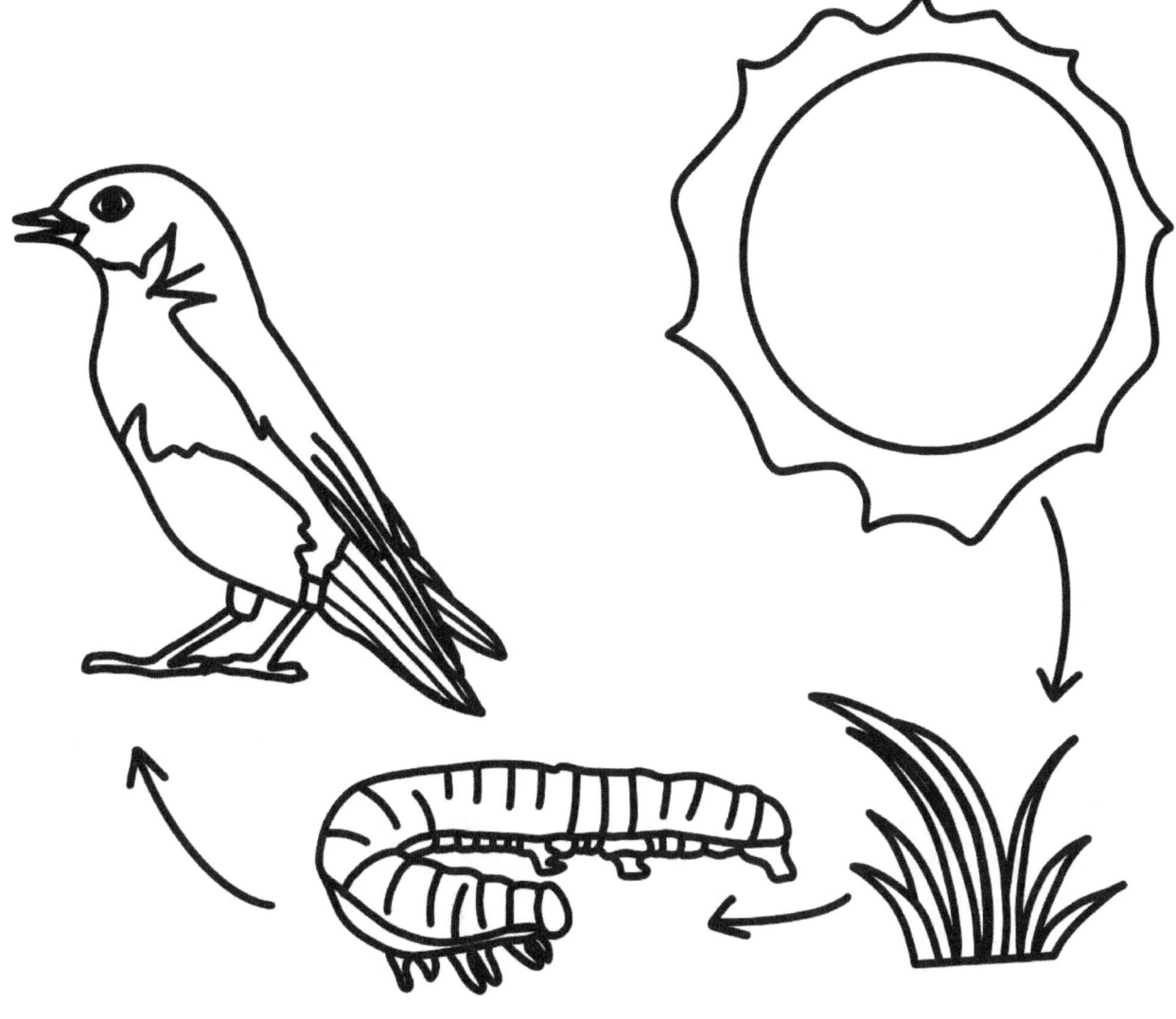

All living things depend on each other for food. A food chain shows you how energy is transferred through organisms.

MONSTERA PLANT

The large leaves on this plant allow it to survive at the bottom of the forest. With little sunlight reaching the forest floor, these large leaves absorb more sunshine,

ADAPTATIONS

The external characteristics of these animals varies depending on where it lives, **how it** moves, **and what it** eats.

Gills to live in the _water_

Teeth to _eat_ coral

Fins to _swim_

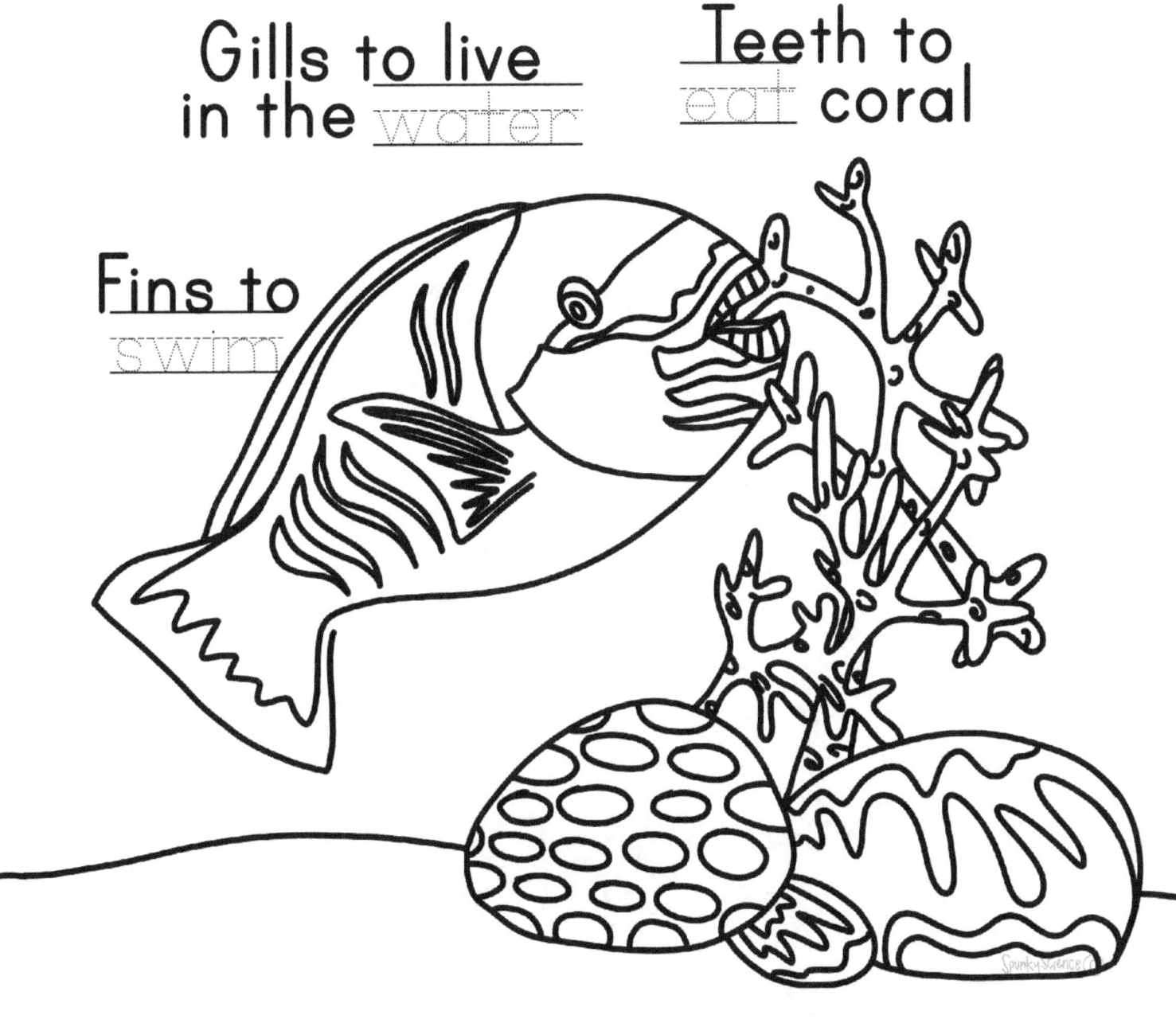

STRAWBERRY PLANT

Young plants and animals have similar features as their parents

PARTS OF PLANTS

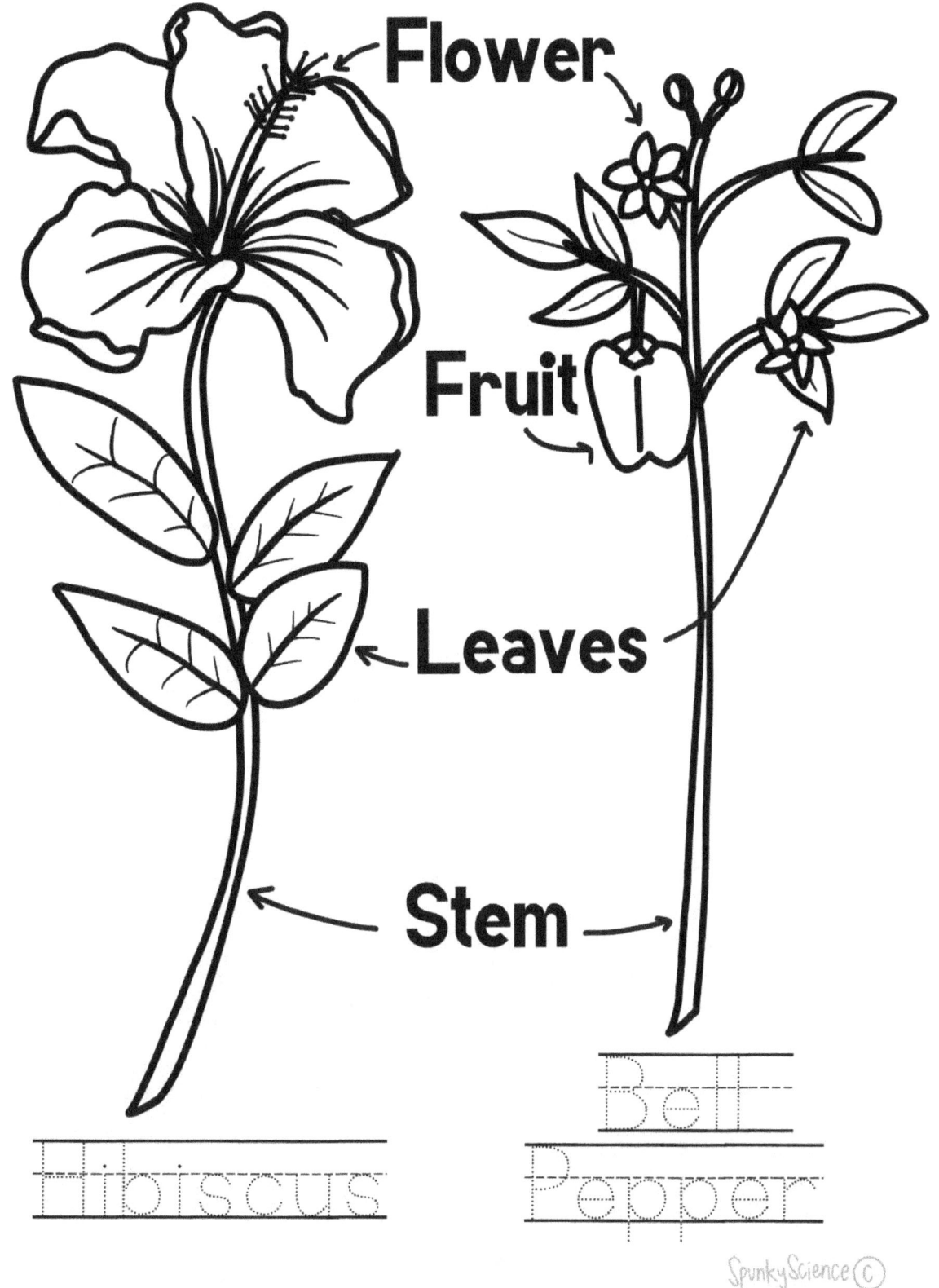

The lifecycle of a
BUTTERFLY

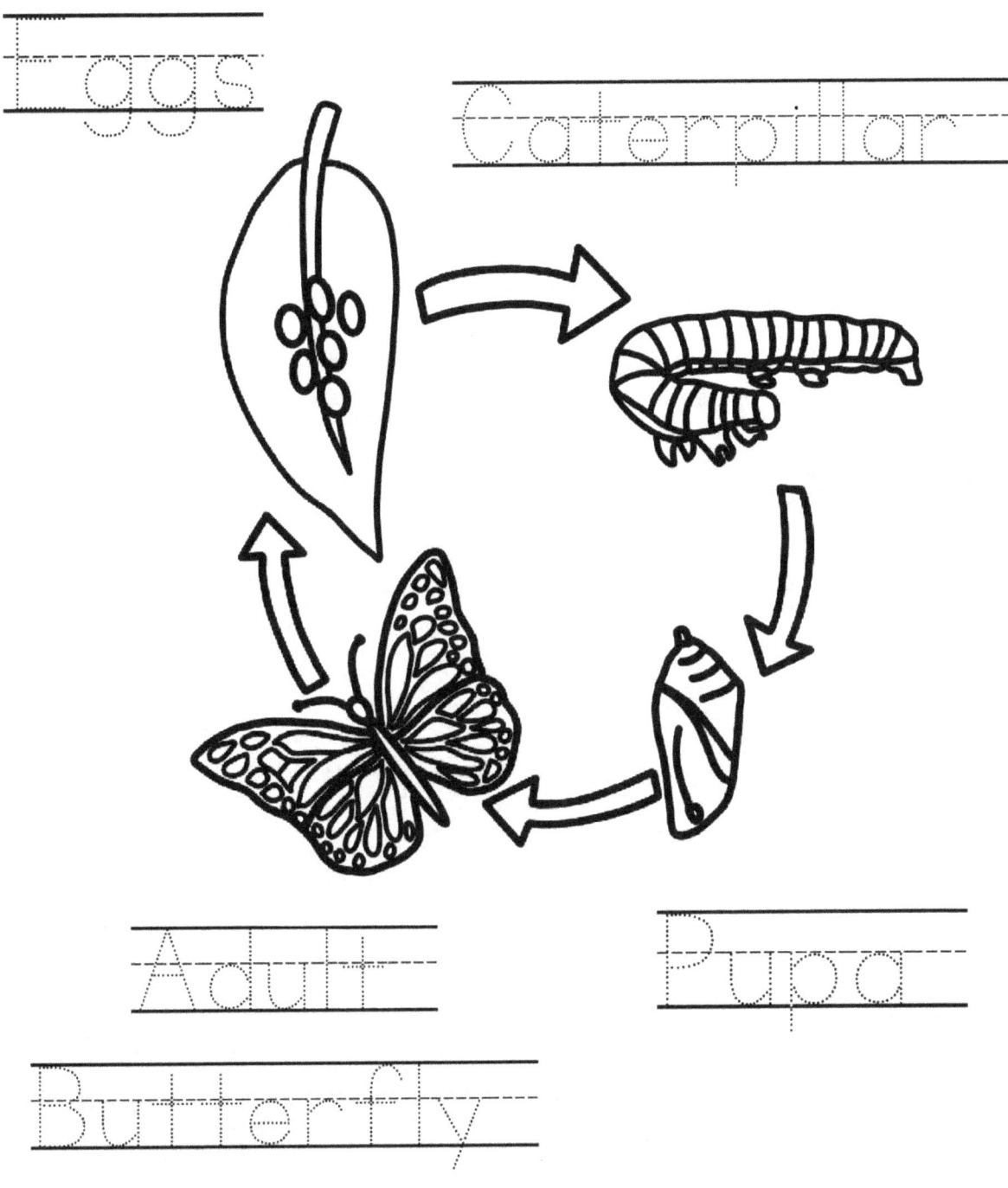

Eggs

Caterpillar

Pupa

Adult
Butterfly

SpunkyScience ©

DOG BREEDS

Breeds of dogs have similar characteristics as their parents

  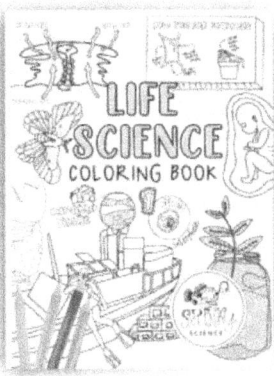 
  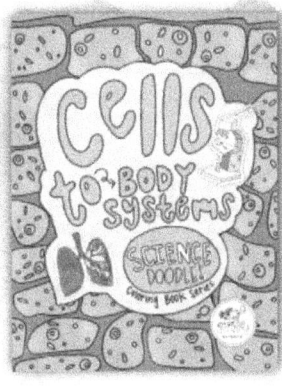 

www.ingramcontent.com/pod-product-compliance
Lightning Source LLC
Chambersburg PA
CBHW060436220526
45465CB00008B/3163